Defining and Analyzing a Business Process

A Six Sigma Pocket Guide

Also available from ASQ Quality Press:

Six Sigma Project Management: A Pocket Guide
Jeffrey N. Lowenthal

Six Sigma for the Office: A Pocket Guide
Roderick A. Munro

Six Sigma for the Shop Floor: A Pocket Guide
Roderick A. Munro

Customer Centered Six Sigma: Linking Customers, Process Improvement, and Financial Results
Earl Naumann and Steven H. Hoisington

The Six Sigma Journey from Art to Science
Larry Walters

Office Kaizen: Transforming Office Operations into a Strategic Competitive Advantage
William Lareau

Reengineering the Organization: A Step-by-Step Approach to Corporate Revitalization
Jeffrey N. Lowenthal

Managing Change: Practical Strategies for Competitive Advantage
Kari Tuominen

Improving Performance through Statistical Thinking
ASQ Statistics Division

To request a complimentary catalog of ASQ Quality Press publications, call 800-248-1946, or visit our Web site at http://qualitypress.asq.org .

Defining and Analyzing a Business Process

A Six Sigma Pocket Guide

Jeffrey N. Lowenthal

ASQ Quality Press
Milwaukee, Wisconsin

Defining and Analyzing a Business Process: A Six Sigma Pocket Guide
Jeffrey N. Lowenthal

Library of Congress Cataloging-in-Publication Data

Lowenthal, Jeffrey N., 1958–
   Defining and analyzing a business process : a six sigma pocket guide /
Jeffrey N. Lowenthal.
      p. cm.
   Includes bibliographical references and index.
   ISBN 0-87389-551-7 (Soft Cover, Spiral Bind : alk. paper)
   1. Process control. I. Title.

TS156.8 .L688 2002
658.5—dc21   2002152027

10 9 8 7 6 5 4 3 2

ISBN 0-87389-551-7

Publisher: William A. Tony
Acquisitions Editor: Annemieke Koudstaal
Project Editor: Paul O'Mara
Production Administrator: Jennifer Gaertner
Special Marketing Representative: David Luth

ASQ Mission: The American Society for Quality advances individual,
organizational, and community excellence worldwide through learning,
quality improvement, and knowledge exchange.

Attention Bookstores, Wholesalers, Schools, and Corporations: ASQ Quality
Press books, videotapes, audiotapes, and software are available at quantity
discounts with bulk purchases for business, educational, or instructional use.
For information, please contact ASQ Quality Press at 800-248-1946, or write to
ASQ Quality Press, P.O. Box 3005, Milwaukee, WI 53201-3005.

To place orders or to request a free copy of the ASQ Quality Press Publications
Catalog, including ASQ membership information, call 800-248-1946. Visit our
Web site at www.asq.org or http://qualitypress.asq.org .

Printed in the United States of America

 Printed on acid-free paper

To my children, Joshua, Julie, Erika, and Gabby. Words cannot express my thanks and love for your strength, support, and love.

A special thanks to my friends at Proforma Corporation.

Contents

Preface

In the body of quality literature, and specifically on the topic of Six Sigma, there is an overwhelming reference to business processes. In just about every article and book written, the author stresses that the reader must clearly define the process that they are exploring if he or she is going to adequately measure and/or change it.

In *Six Sigma Project Management: A Pocket Guide* (ASQ Quality Press), I state that "Six Sigma focuses on two things: the customer's requirements and the processes meant to fulfill those requirements" (p. xii). Here rests the problem. When attempting to define a process, most practitioners use standard flowcharting. By using flowcharting, however, one leaves out critical information necessary for change. That is, flowcharting shows decision points. It does not show departmental interactions nor communication patterns. This is where this pocket guide fits in. To take a detailed look at your business processes, you will require additional tools. This pocket guide provides you with a series of "other" visual tools that will give you a big-picture view of your process.

The first section will, very briefly, outline what a process is. There are many books on the market that will provide you with greater detail about process definition than this pocket guide. This guide's primary focus is mapping tools.

The next two sections will outline two visual tools—Business Interaction Diagrams and an Integrated Flow Diagram. The final section describes a method to define change to your process after you have mapped it out. Together, all four sections will provide you with the tools necessary to make effective and profitable change.

In each of the sections, I will outline how the various tools are used. In addition, I will provide a case study example of the use of that specific tool.

In closing, I have attempted to make a practical guide, one that is easy to read and can be applied immediately. The tools are not hard to use but are very powerful. Have fun on your exploratory journey in defining and analyzing your processes.

Jeff Lowenthal, PhD
West Bloomfield, Michigan

Chapter 1

Introduction

This guide is about the analysis and design of work-flows and processes within and between organizations. As you start working on a Six Sigma project, one of the first critical steps is to define the process that you are exploring. This guide will also help you define the process of identifying project boundaries.

WHAT IS A BUSINESS PROCESS?

A process is a structured, measured set of activities designed to produce a specified output for a particular customer or market. It implies a strong emphasis on how work is done within an organization. A business process is a set of logically related tasks performed to achieve a defined business outcome.

Two important characteristics of processes are that they:

1. Have customers (internal or external).

2. Cross organizational boundaries, that is, they occur across or between organizational subunits.

Processes are generally identified in terms of beginning and end points, interfaces, and organization units involved, particularly the customer unit. High-impact processes should have process owners. Examples of business processes include: developing a new product; ordering goods from a supplier; creating a marketing plan; processing and paying an insurance claim; and so on.

Processes may be defined based on three dimensions:

1. *Entities.* Processes take place between organizational entities: interorganizational, interfunctional, or interpersonal.

2. *Objects.* Processes result in manipulation of physical or informational objects.

3. *Activities.* Processes could involve two types of activities: managerial (for example, develop a budget) and operational (for example, fill a customer order).

BUSINESS PROCESS AND SIX SIGMA

Work within organizations is completed by a process or various processes. A child's plastic toy is usually created with a process of injected plastic. Issuance of a purchase order is completed with a transaction process. Because business processes contain several major steps, there are opportunities for breakdowns. That is, steps may not be completed the same way each time the process is initiated. This nonconsistency is termed *variation.*

One of the principal goals of a Six Sigma effort within a company is to reduce process variation, that is, to develop an approach to limit the variation and more tightly focus the process so as to produce the same results over a long period of time. To reduce this process variation, you need a clear understanding of what the process is and how it works. Mapping a process helps the Six Sigma professional to identify the flow of events in the process as well as the inputs (X's) and outputs (Y's) in each step. The easy part is defining what goes into a process and the desired results. The hard part is trying to figure out the variables between input and output, which are also called *functions.* Graphically, you can display a process as seen in Figure 1.1.

For example, if we looked at the variation in production throughput we would find many functions or variables that impact the end result. Graphically, it would look like Figure 1.2.

When you map the process that you are investigating, the visual diagram created shows all the major and

Figure 1.1   Graphical display of a process.

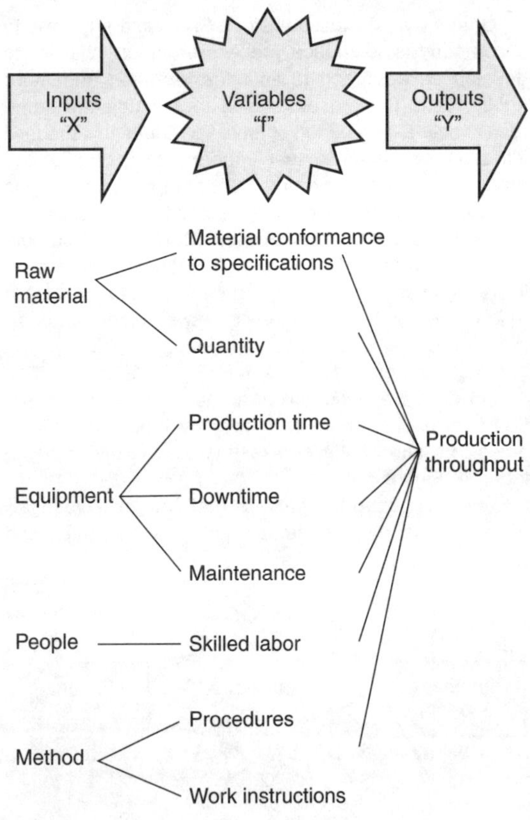

Figure 1.2   Production functions.

minor steps of the process. Further, it identifies key input variables and the resulting output. Finally, it tends to show the complexity of the process by displaying value-added and non-value-added activity. In addition, a process map is an effective communication tool. It ensures that all Six Sigma team members view the process in the same fashion.

Thus, if your one of your goals of a Six Sigma project is to reduce process variation, the tools in this pocket guide will help you graphically represent the various functions or variables within your business process. This will greatly aid in determining which of these functions are having the greatest impact on your output and aid you in facilitating change.

SIMPLE CASE STUDY

To help you better understand the methods outlined in this pocket guide, a fictitious company (based on a real-life example) was created—Premier Folding Kayaks. Premier Folding Kayaks have been around for about 40 years. What makes them so popular is that they fold into a small bag for easy portability. Recently, executive management decided that they wanted to significantly improve their production process. They embarked on a change effort using the approach outlined in this pocket guide.

SUMMARY

In a review of the literature regarding Six Sigma, in a majority of cases there is a reference to "mapping the process." The purpose of this activity is to increase the communcations between team members by ensuring that everyone is viewing the process the same way. Further, it keeps management informed as to the area that the team is exploring. The problem, however, is the depth that is offered on the topic. Most resources, if not all, only provide about a paragraph or two on process modeling. In very few cases, you might get an entire half of a page. This is like informing a carpenter that they need to use a screwdriver, but forgetting to tell them the type and size needed.

This pocket guide outlines all the tools needed to complete a detailed process mapping effort. As a practitioner, use those graphic tools that you deem appropriate to the effort, based on your resources and scope of work. Let's move on to the first visual tool—Business Interaction Diagrams.

Chapter 2

Business Interaction Diagrams

This section provides the first systemic approach to learning more about the process that you are investigating. Specifically, you will learn the steps necessary to examine the key operating functions of the process. Understanding these elements is critical for improvement of the process.

FOUNDATION STEPS

Every process is comprised of a series of functions, or events. The goal of Business Interaction Diagrams is to identify the key events within the process, then lay out these events in a rational manner according to the sequence of events.

Step 1: Identify the Major Operating Events, or Functions

1. The major events are defined as those activities about which the firm needs (or wants) information and are strategically important to the firm.

2. There are three relevant types, or categories, of events:

   a. *Operating events*—Activities that enable the firm to conduct its business. For example, taking orders, picking inventory, shipping products, receiving merchandise from vendors, receiving payments, and so on.

   b. *Decision/Management events*—Activities in which management or others plan, control, or evaluate a process. For example, deciding to add a new product line, undertake a marketing effort, or make-or-buy decisions. Typically, information about these are not collected or maintained.

   c. *Information events*—Events that take one of three forms, and are triggered by operating and decision events:

      - *Recording*—Events that involve recording or capturing data about something and storing it, for example, recording an order. (Note that the operating event of receiving a customer order "triggers" the information event involving recording the order.)

      - *Maintaining*—Events that involve keeping data current, for example, changing employees' addresses, updating customer information, and deleting discontinued product information from the database.

- *Reporting*—Events that involve providing information to those who need or request it. For example, providing financial statements, managerial reports, performance reports, reports on cost information, and so on.

Step 2: Determine the Resources, Agents, and Locations

- *Resources* are items that have economic value to the firm, for example, cash, inventory, equipment, supplies, warehouses, factories, land, and so on. To determine the resources (if any) involved in each event from step 1, you should ask, "What things of value are being used or are involved in this event?"

Do not get carried away by identifying every possible resource. You are trying to identify just the major resources involved, about which the firm wants information. For example, if you are analyzing the purchase of groceries at a grocery store, then the groceries purchased and the cash received are resources. You do not necessarily need to identify the cash register involved, the utilities needed to run the register or power the lights, or the plastic sacks. In these cases, even though they are resources, the company probably doesn't want or need information about them at this time.

- *Agents* are the people or organizations that participate in the events. They may be internal (that is, work for the firm) or external (that is, suppliers or customers). It is usually relatively easy to identify agents. There is always at least one internal agent, and most of the time an external agent, involved in each event.

• *Locations* should only be identified and included to the extent that they affect the event in some way. It is important to keep track of, or collect, information about a location because that information changes over time or is vital to the business.

For example, a health club will want to collect information about racquetball court reservations (this changes all the time) but will probably not care to collect information about the courts themselves (this does not change). An airline will need information about seat assignments, gates, and departure and arrival times because these are vital to the business, but they do not need or care about the location of the operator who reserved your flight.

Step 3: Document the Relationships That Exist

• One method to complete this task is to use a spreadsheet. Place each item (resource, agent, and so on) in a separate column. Connect each resource, entity, internal agent, external agent, and location associated with that event either by drawing (a) line(s) between the columns or creating some other marking system.

• Identify the nature of the relationship somewhere on the line or in a key based on this other marking system.

• Connect the related events. This connection represents (and is caused by) the "give and get" nature of the events. For example, you would want to connect a sales event with a cash-collection event.

MAPPING THE PROCESS

This method of mapping (see Figure 2.1) actually consists of three interrelated models or diagrams (one has several subsets):

- Hierarchy models
- Business interaction model
- Workflow model

The following pages provide details of the various models.

Hierarchy Models

Hierarchy models will enable you to develop visual representation of "support" functions in your organization. The three diagrams that you will create are:

1. Organizations

2. Goals

3. Locations

The hierarchy models organize business objects in a top-to-bottom structure or *hierarchy*. Using hierarchies, you can capture successive levels of detail in your business entities, roles, business processes, and other business objects. Analyzing your business from top to bottom (or bottom to top) provides an efficient way of capturing all of the detailed objects that comprise larger objects, enabling you to obtain comprehensive macro and micro views.

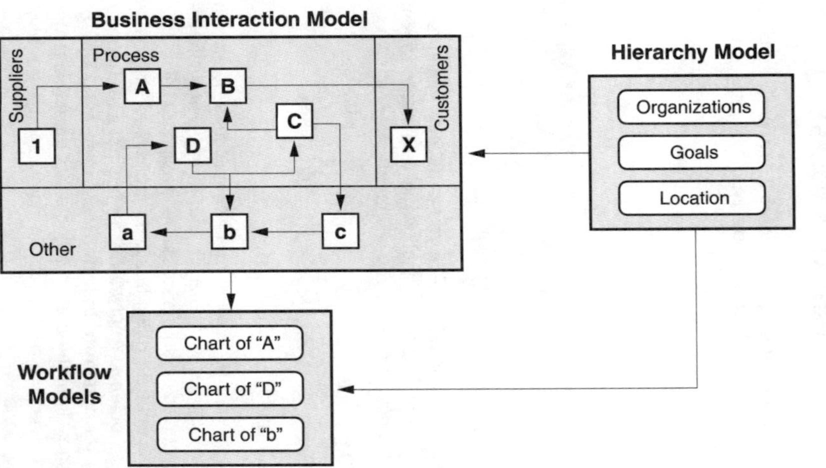

Figure 2.1   Interrelated models.

Organization Model

An *organization model* is based on any of the following object types:

Markets	Business entities that interact with the business, either as suppliers or customers
Organizations	The complete units within a business
Roles	Jobs or roles performed within an organization

An *organization* is a group of people that make up a business or interact with it. Markets, organizations, and roles are organizations. An organization hierarchy places either a market or an organization at the highest level in the model. More specific organizations within the general one are positioned below it. The lower in the hierarchy model the organization appears, the more detailed the organization.

In Figure 2.2, the Premier Folding Kayaks company is broken down into its composite roles.

Goal Model

A *goal model* is a hierarchy model that organizes the goals set for the business. A goal model places a general goal or goal category at the highest level. More specific goals are organized below it to support or further describe the top level goal identified. The lower the goal appears in the hierarchy, the more detailed it is.

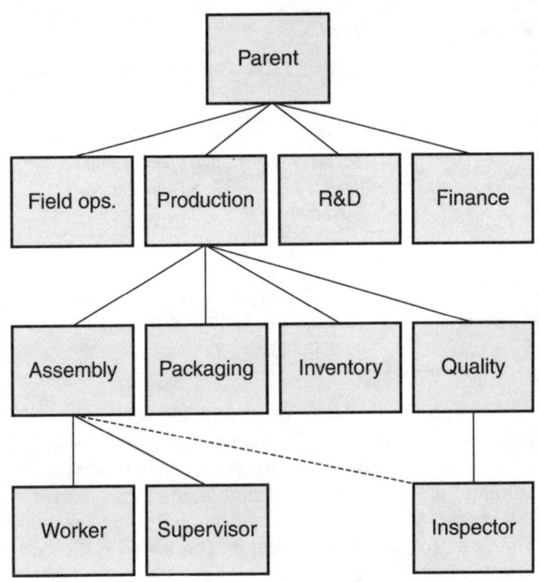

Figure 2.2 Composite roles of an organization.

In Figure 2.3, Premier Folding Kayaks identified four key goal categories. Only the operational goals are further displayed in Figure 2.3.

Location Model

A *location model* is a hierarchy model that organizes the locations for a business. A location model places a general location or location category at the highest level in

Figure 2.3   Operational goals of an organization.

the model (that is, corporate HQ). More specific locations within the general location are organized below it. The lower the location appears in the hierarchy, the more specific the location.

In Figure 2.4, the key location *Sales Regions* is broken down into the two composite locations *North America* and *Europe.*

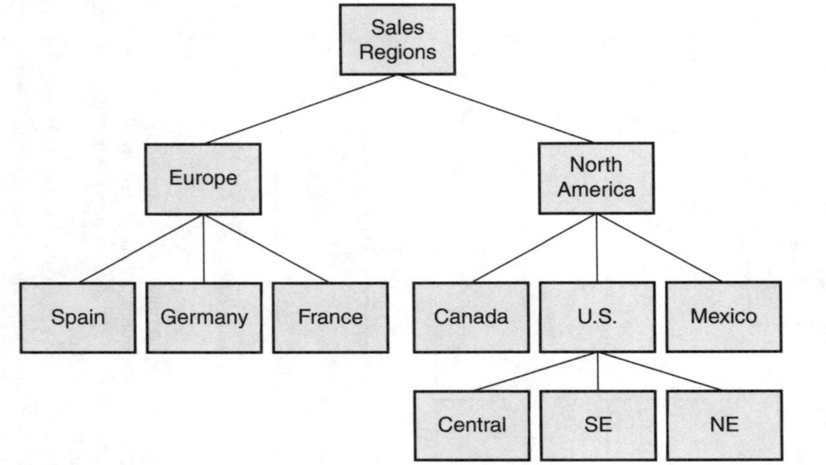

Figure 2.4   Sales regions for North America and Europe.

Business Interaction Model

This model provides a quick way to develop a "macro" view of a business with respect to its customers, suppliers, competitors, and major organizational interfaces. Business interaction models are partitioned into four key sections: suppliers, competitors, customers, and the business domain. Figure 2.5 shows the format of this model.

Step 1: Define Departments

List all of the key departments involved with the process or domain that you are investigating. Place them in the center area of the diagram and make sure you also title each block in the diagram (see Figure 2.6).

Do not link the boxes yet. This will be completed in step 3.

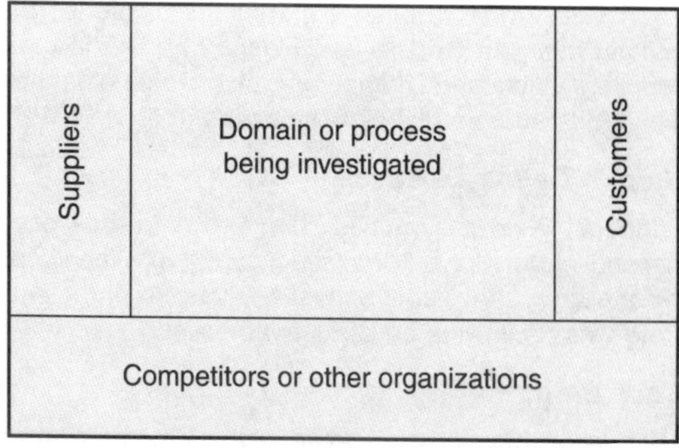

Figure 2.5 Format of the business interaction model.

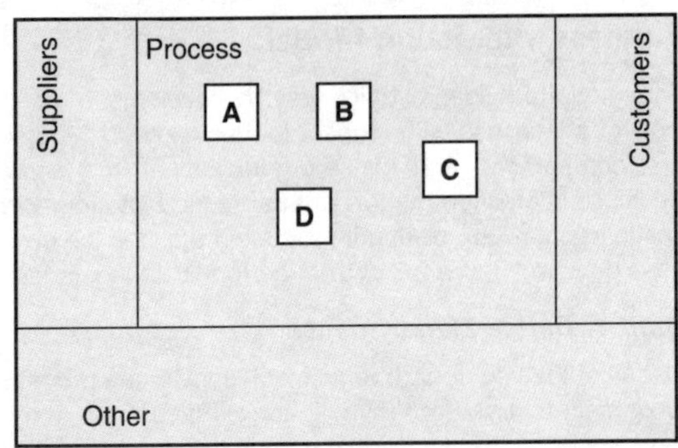

Figure 2.6 Defining departments.

Step 2: Define Other Entities

As in step 1, list all other departments or groups that interact with your process (see Figure 2.7). This includes suppliers, customers, computer systems, and other entities. Make sure you label these boxes, too.

Step 3: Define Linkages

Connect the various entities using arrows to show physical and informational flows (see Figure 2.8). When labeling the arrow, the label should be concise and clearly state what is flowing between the blocks.

Case Study

After an examination of Premier Folding Kayaks, nine primary organizations or departments were identified. Talking with each one, inputs and outputs to the organization

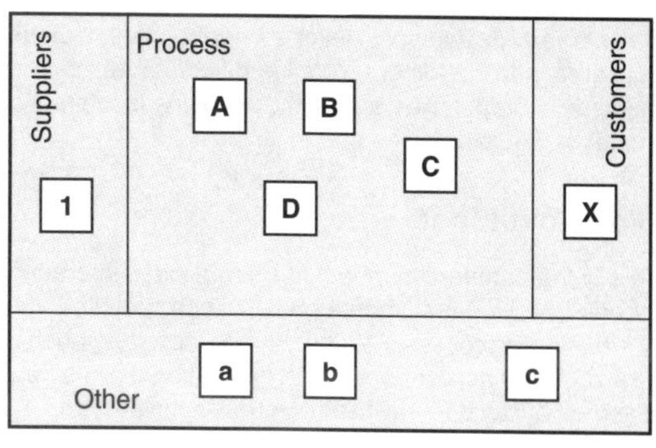

Figure 2.7   Defining others who interact with your process.

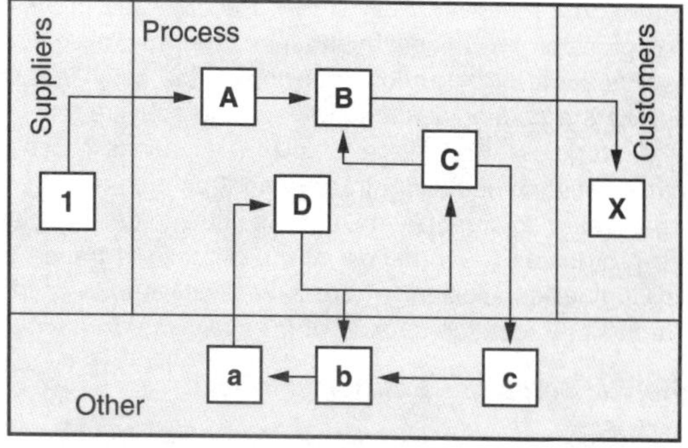

Figure 2.8   Making the linkages.

were noted. In addition, several support organizations, suppliers, and customers were identified, as well as their inputs and outputs. As a result, the business interaction model in Figure 2.9 was created.

Workflow Model

This is the companion model to the business interaction model. The workflow model gives a more detailed view of a business process, helping you visualize and analyze how multiple departments or organizational units work together by evaluating their internal activities and the passing of deliverables among them.

A *workflow model* is a representation of a business process in terms of its component activities and the flow of work between the activities. A workflow model concentrates on the flow of work through the business for a single output (that is, a product or service) or a single input (that is, the handling of an order). Because the process may cross organizational boundaries, the workflow model depicts these organizations performing the activities, as well as the communication between the activities.

A workflow model is a map of a business process starting with its initiation, tracing work as it passes from organization to organization, until its ultimate deliverables are produced. The workflow of the process is traced to its ultimate conclusion, where a final deliverable is forwarded to a location outside of the business domain.

Step 1: Select an Entity to Model

Using the business interaction model just developed, select one or more functions to define further, as seen in Figure 2.10.

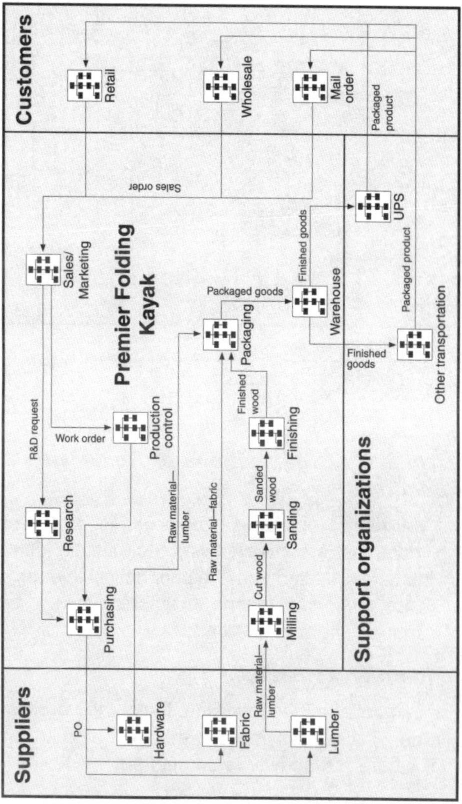

Figure 2.9 Business interaction model.

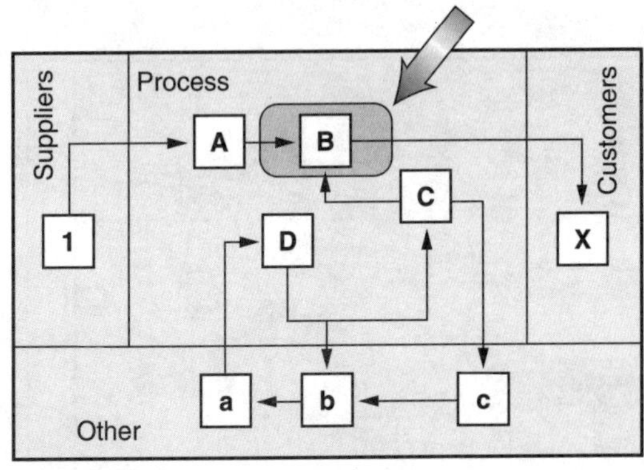

Figure 2.10   Defining a function.

Step 2: Identify Organizations or Roles within That Function

Now that you have identified a focal point, list all organizations or roles that are involved with the function identified. Use the data collected from the prior defined hierarchy models. Place the organizations in different rows on a sheet of paper as shown in Figure 2.11.

Step 3: Identify Activities

Using the symbols defined following, lay out the process step by step. Place the symbol in the row (also known as a lane) where the activity is completed.

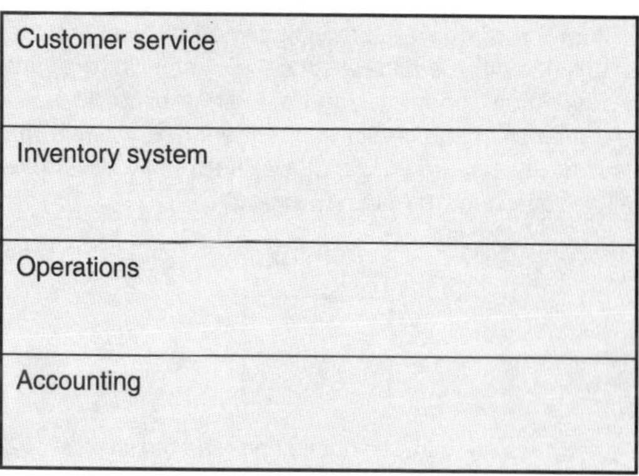

Figure 2.11   Listing of roles involved with a function.

Workflow Symbols

When completing this specific model, you should use traditional or standard practice flowchart symbols. Below is a description of several symbols that are often seen in this model. Please note that this is not an exclusive set of symbols, just the more frequently used icons.

Store

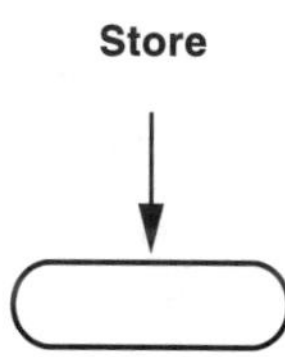

A group of deliverables maintained by, and provided to, activities in a business process. A store may appear in a workflow model as the destination of workflows to hold their deliverables for subsequent use. A store may also be the source of workflows that pass deliverables which have been previously stored.

Activity

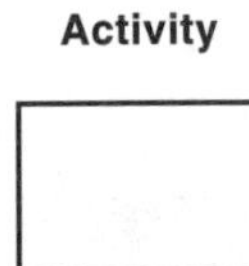

Activities are initiated based upon the occurrence of an event. This event may be the completion of a previous activity, the arrival of a deliverable from outside the business process, or a condition that occurred within the business domain. In response to the event, the activity takes the input(s) coming from (a) previous activity(ies) and transforms it/them into one or more deliverables.

Source or Sink

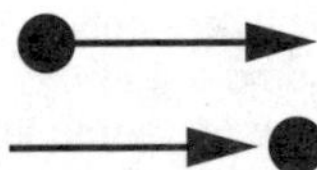

The upper symbol is that of a source, an entry point to a workflow model. A source is used to display a deliverable received by an activity whose previous activity or origin is unknown. If the entry point to this model is a previous activity or of known origin, an activity box is used. The lower symbol shows a sink. This symbol is used to display a deliverable sent to an activity or some other unknown destination.

Decision Point

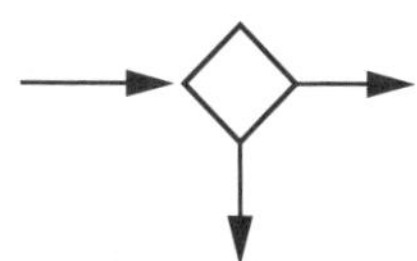

The flow of work through the business process may not be uniform each time, but can vary based upon conditions. A decision point evaluates a specified condition during the processing, and directs the flow of work based upon its evaluation. A decision point receives a single workflow and has multiple possible paths to subsequent activities based upon the number of alternatives for the condition.

Junction

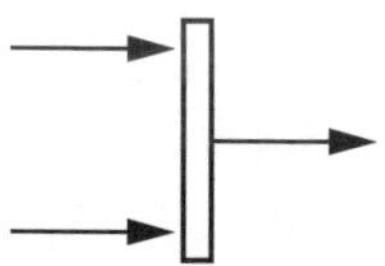

This symbol allows a workflow (or workflows) to converge/diverge into another workflow(s). Junctions are used to combine or split workflow deliverables as they move from one level in the process hierarchy to another.

Bridge

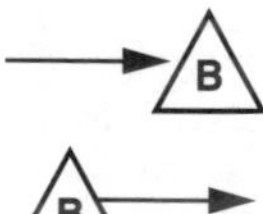

This symbol allows a workflow to be placed in a model without the physical line stretching between the two activities. A bridge may be thought of as a modeling convenience for looping to a previous activity, or for reducing the need for long workflow lines. The bridge operates by placing a reference point on the workflow's origination and destination.

Step 4: Link and Label Symbols

In this final step, you link all the symbols displayed in the rows or lanes with arrows, as shown in Figure 2.12. Next, you label each of the arrows with the data, information, or physical product that is moving from block to block (not shown in Figure 2.12).

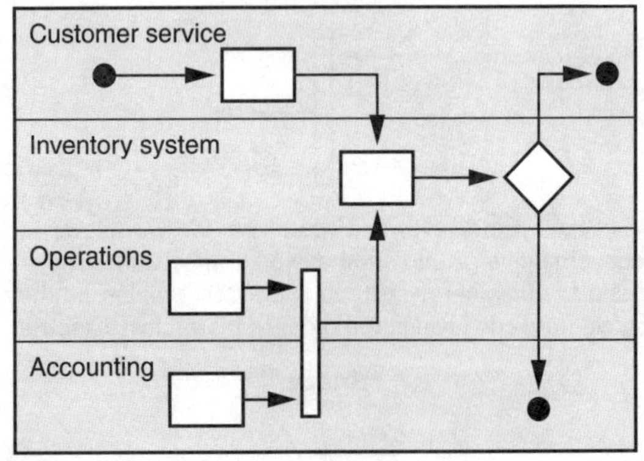

Figure 2.12   Linking the symbols.

HOW TO DRAW THESE MODELS

The easiest method for creating these models is to use flip-chart paper and Post-It notes. Use the figures presented earlier as your guide to laying out the paper. Next, draw the various symbols on Post-It notes and place them in the correct areas. To draw arrows, some people write on the paper itself; others have used string or yarn. Pick a method that works for you.

CASE STUDY

In a review of the business interaction model and after talking with several employees, Premier Folding Kayaks' process modeling team decided to start its detailed analysis in the packaging area (see Figure 2.13). Its rationale was that it was the central point prior to shipping product to the warehouse, that is, a back-to-front approach.

Figure 2.14 presents the detailed steps (and supporting activity) for the packaging area.

SUMMARY

The purpose of this chapter was to outline the first of two visual tools—the Business Interaction Diagrams. This tool consists of several submodels:

1. *Hierarchy models* enable you to develop composition structures for the following business objects:

   - Organizations

   - Locations

   - Goals

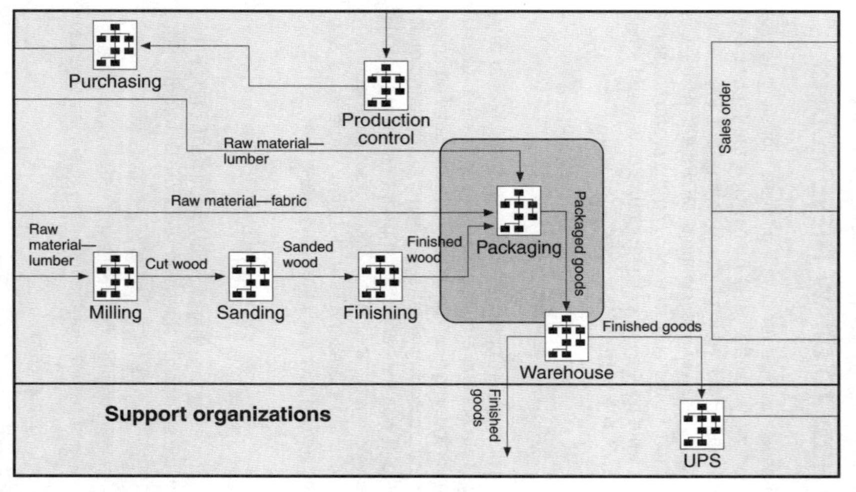

Figure 2.13   Detailed analysis of packaging area.

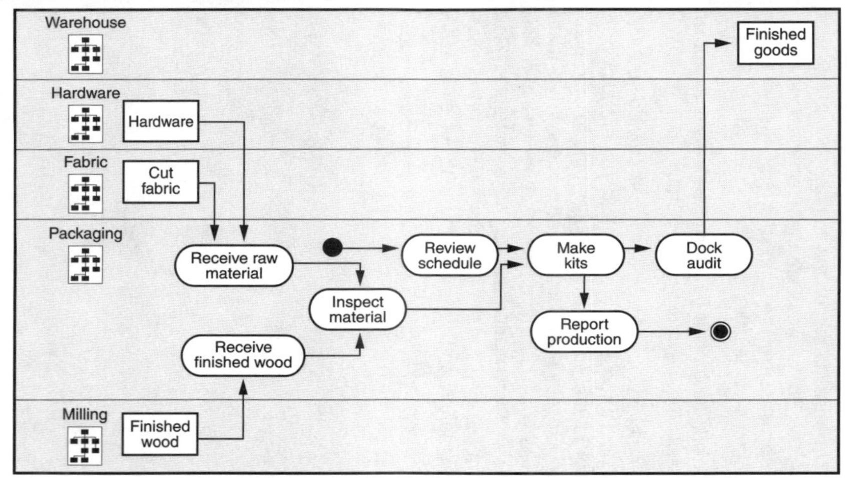

Figure 2.14 Detailed steps for packaging area.

2. *Business interaction models* are divided into four areas. These areas are the business domain or process, suppliers, competitors or other organizations, and customers.

3. *Workflow models* outline in detail the activities or tasks that need to be completed.

In combination, these models will provide you with lots of detail about the process that you are exploring. One key failure point is that people tend to stop mapping the process after using these models. While these models provide good information about your process, there is still information missing. The Integrated Flow Diagram presented in the next chapter fills in this missing information and provides you with an even more detailed picture of the process.

Chapter 3

Integrated Flow Diagramming

In this chapter, you will be introduced to a second visual tool—the integrated flow diagram.

OVERVIEW TO STRUCTURED ANALYSIS

This chapter is about structured analysis, which is a new way of looking at a process. To help explain, let's look at the following example.

The following was taken from Premier's mountain folding kayak's instruction manual.

Assembly Instructions for Folding Boats

1. *Lay out hull in grass (or on carpet). Select clean, level spot.*

2. *Take folded bow section (with red dot), lay it in grass, unfold 4 hinged gunwale boards. Kneel down, spread structure lightly with left hand near*

bow, placing right hand on pullplate at bottom of hinged rib, and set up rib gently by pulling toward center of boat. Deckbar has a tongue-like fitting underneath which will connect with fitting on top of rib if you lift deckbar lightly, guide tongue to rib, press down on deckbar near bow to lock securely. Now lift whole bow section using both arms wrap-around style (to keep gunwales from flopping down) and slide into front of hull. Center seam of blue deck should rest on top of deckbar.

3. *Take folded stern section (blue dot, 4 "horseshoes" attached), unfold 4 gunwales, set up rib by pulling on pullplate at bottom of rib. Deckbar locks to top of rib from the side by slipping a snaplock over a tongue attached to top of rib . . .*

And so forth.

A STRUCTURED LOOK

An integrated flow diagram (IFD) is a graphic representation of the physical and communication patterns of a process. Remember that the business interaction models show the interrelationship of departments and the tasks to be completed. Flowcharts show decision points. With this tool, you are only interested in physical and communication flow. Figure 3.1 shows a graphic representation of Premier's instructions presented above.

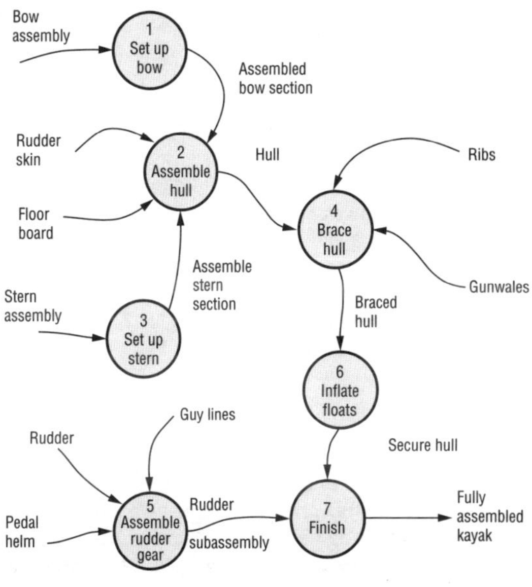

Figure 3.1   Graphical display of kayak assembly.

INTEGRATED FLOW DIAGRAM CONVENTIONS

An IFD is made up of four basic elements—pipelines, activities, files, and external entities. The symbols and method used to complete an IFD are adapted from a system-analysis technique called *data flow diagramming.* The principal differences between the IFD and the data flow diagram involve their application and scope. For example, data flow diagrams were designed for data processing applications, whereas the IFD can be used for any business process. Further, while both data flow diagrams and IFDs show information flows, IFDs also display the flow of physical products through the process.

Figure 3.2 is a simple example of an IFD. In this example, X arrives from the external entity S and is transformed into part Y by activity P1. Activity P1 requires some type of information or item from file F. Y is subsequently transformed into Z by activity P2.

The First Element: Pipeline

A *pipeline* moves a single package of information or material between the activities, files, and external entities shown on the IFD. It is symbolized by a named arrow to show the interface as shown in Figures 3.3 and 3.4.

Below is a useful set of conventions for dealing with pipelines:

- No two pipelines have the same name.

- Names are chosen to represent not only the package that moves over the pipeline, but also what is known about the package.

- Pipelines that move into and out of files do not require names; the file name will suffice to describe the pipeline. All other pipelines must be named.

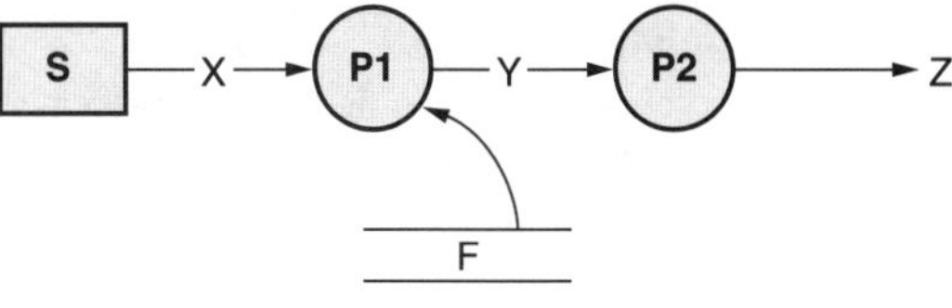

Figure 3.2 A simple IFD.

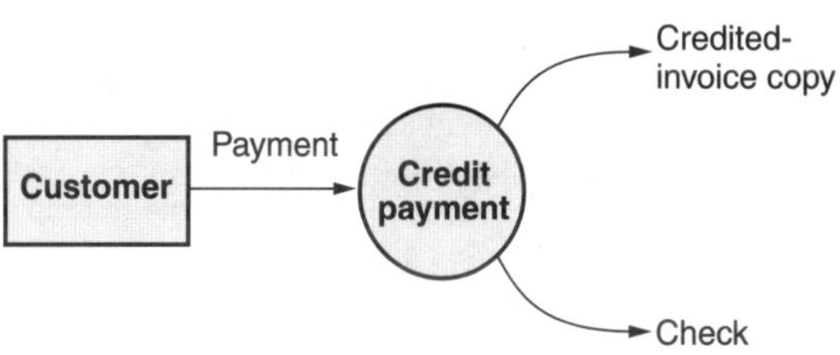

Figure 3.3 Example of a physical pipeline.

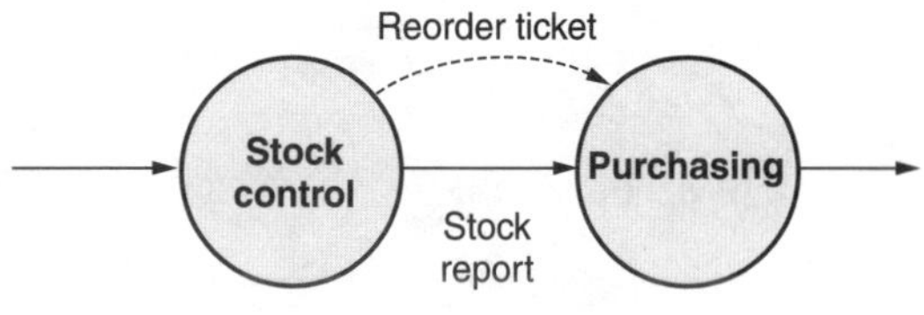

Figure 3.4 Example of a verbal pipeline.

Types of Pipelines

Thin, thick, and broken lines can be used to distinguish between types of pipelines, but the IFD's creator may apply a variety of conventions to these. The thick-line arrow (or single head arrow) often represents a flow of physical materials from point to point, such as the movement of wood boards from a storage facility to a cutting area. In turn, the thin-line arrow (or double head arrow) is often used to represent the flow of information—such as reports or telephone calls—between activities. However, some IFD creators use the thick line to show the principal flow, or main manufacturing pipelines, while using thin lines to represent support pipelines, for example, scheduling and production control.

In Figure 3.5, the solid pipeline represents the main or primary flow. The dotted pipeline shows a secondary flow. The third pipeline is verbal in nature and is represented by a double arrow head.

Two other points are worth noting here. First, you should be careful not to attempt to represent pipeline control points on the IFD, as such items should be represented

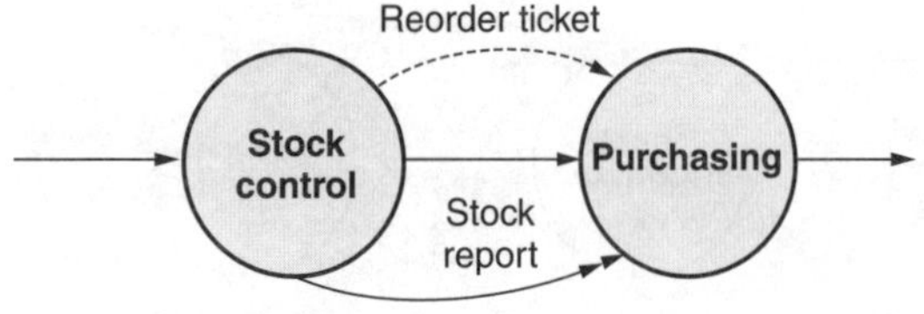

Figure 3.5   Types of pipelines.

on the flowchart. Second, if a pipeline cannot be named easily, it is probably not defined adequately. Either several packages are being transmitted along the pipeline, or an incomplete package is being transmitted.

The Second Element: Activity

Activities represent some amount of work performed on a package. A common convention is to represent activities by circles on the IFD, though ovals are sometimes used. Either way, make sure to give the represented activity a name that clearly and accurately describes the activity. Each activity should be numbered, with the numbering convention depending on how the various diagrams interrelate. See Figure 3.6.

Sometimes several pipelines must be present in order for the process to do its work. At other times, the process or activity only needs the input from one of several pipelines. On the other hand, the output can travel out one

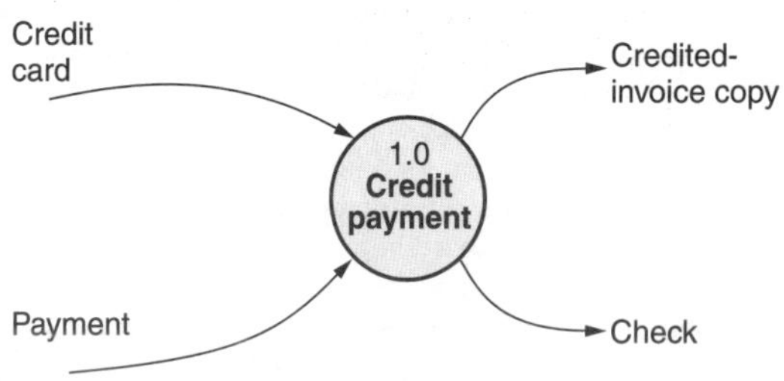

Figure 3.6   Name and number the activity.

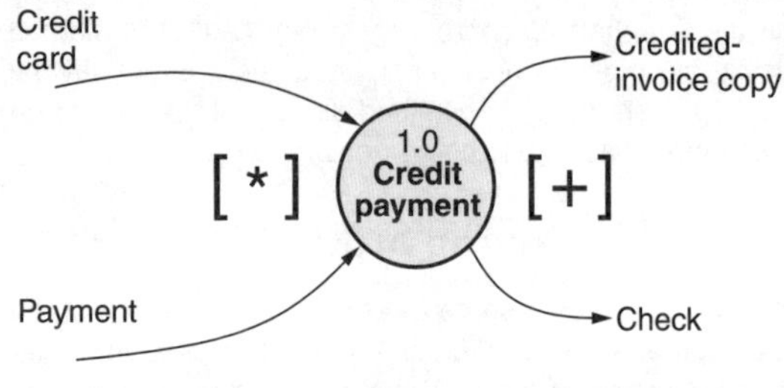

Figure 3.7   Symbols representing distinct paths.

or another pipeline or both. The following symbols are used to represent these distinct paths (see Figure 3.7):

- [*] means *and.* It portrays a conjunction of pipelines.

- [+] means *or.* It portrays a disjunction of pipelines.

The Third Element: The File

A *file* is a collection of information or material, or a space where this information or material is stored. For an IFD, a file is any temporary repository for information or material. Examples of files for information are computer tapes, a specific area on a computer disk, a card data set, an index file, or an address book. Files for material include, for example, loading docks, warehouses, racks in warehouses, and even wastebaskets (the circular file). Two parallel lines often represent a file, with the file's name in close proximity. Since the IFD must be meaningful to its users, file names also need to be clear and meaningful. Avoid using coded names or abbreviations for files, and be sure not to give the same name to two files on a single IFD.

The Final Element: External Entities

Any process can be described on an IFD with pipelines, activities, and files. However, an IFD is clarified and becomes more useful when the process is shown in the larger context of external entities. An *external entity* is a person or organization outside the boundary of the process that is a net originator or receiver of the process being mapped. The key qualifier here is the phrase "outside the boundary of the process." A person or organization inside the boundaries is characterized by an activity on the IFD.

By convention, external entities are represented by named boxes, with pipelines flowing into and/or out of a single box (see Figure 3.8). Boxes, however, should be

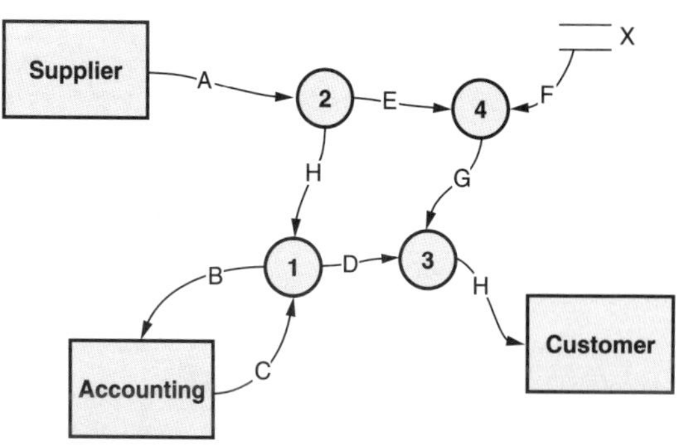

Figure 3.8   External entities of pipelines.

used sparingly on the IFD, and they should not represent major concerns. External entity boxes exist only to provide commentary about the process's connection to the outside world. If a box represents a major concern in an IFD, the boundaries of the process are probably not defined correctly.

LEVELED INTEGRATED FLOW DIAGRAM

Sometimes you can draw an IFD on a single sheet of paper. Other times, you can find yourself working on a table-size sheet of paper with hundreds of activities. When your IFD is too large to be shown on a single page, you need to partition the process into subsystems. The highest level uses the numbering convention of 1.0, 2.0, and so on, for the various activities. To "explode" or dig deeper into a specific activity, you would number the next level activities as 1.1, 1.2, 1.3, and so on. This leveling convention can be seen in Figure 3.9. Please note that the final level of your IDF should be a flowchart (see Figure 3.10).

GUIDELINES FOR DRAWING IFDs

When attempting to draw an IFD for a particular process, the following sequence is recommended:

1. Identify net input and output pipelines. Draw the input and output pipelines around the outside of the diagram.

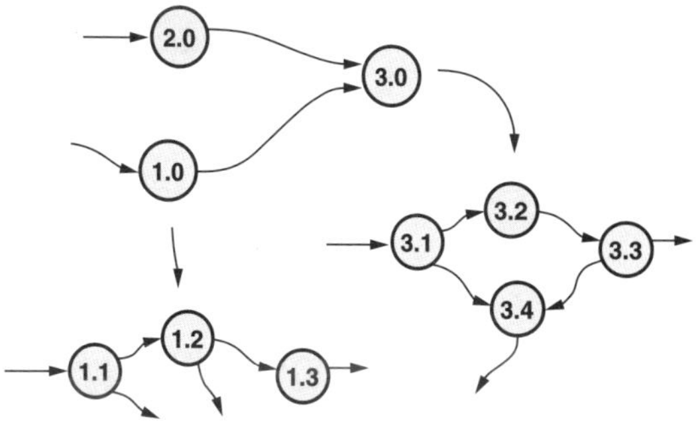

Figure 3.9 High-level numbering convention.

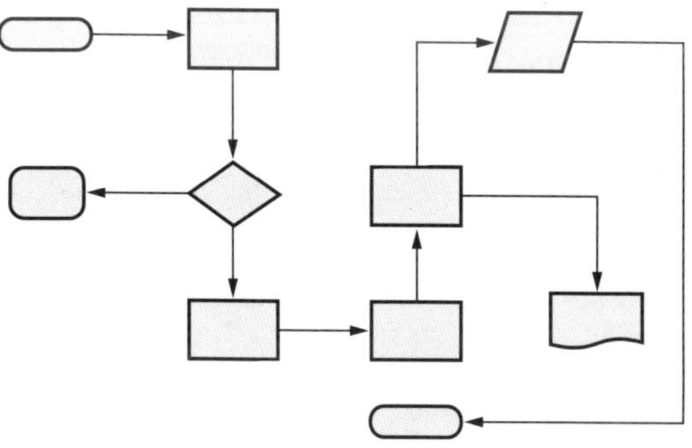

Figure 3.10 Final level of IFD—a flowchart.

2. Fill in the body of the IFD and name the activities.

3. Label the pipelines.

4. Be prepared to start over.

In the following sections, each of these five steps will be presented in greater detail.

1. Identify Net Input and Output Pipelines

The business of deciding net inputs and outputs is closely tied to the decision about the boundaries of the project. Selecting the initial boundary is a matter of judgment and feel. The team needs to select a boundary that is large enough to include all relevant activities to the effort, but small enough to exclude irrelevant activities.

After defining the boundaries of the project, examine the pipelines that cross those boundaries. These pipelines are the net inputs and outputs. List the net inputs and outputs on the periphery of the diagram.

Two notes: First, use a broken line to define net input and output boundaries. Within this convention, the boundaries can be identified with a quick glance at the IFD. Second, don't be too worried about being complete at this point—the process of developing an IFD has self-checking mechanisms that identify forgotten pipelines. See Figure 3.11.

2. Fill in the Body of the IFD and Name the Activities

The first step is to concentrate on activities. In particular, look for any major activities in the process. For example,

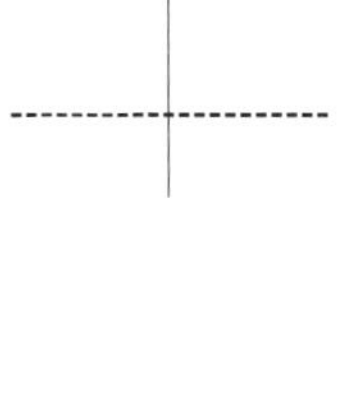

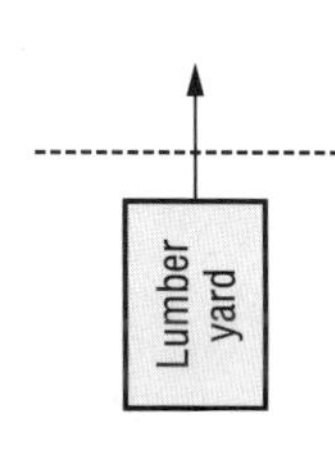

Figure 3.11   Net input and output boundaries.

in the production process at Premier Folding Kayaks, there are four major activities linked by pipelines that represent the primary flow of material through the production process (see Figure 3.12). Thus, the team would enter the major activities on the diagram and connect them with the pipelines on the periphery. Name these activities at this time (see Figures 3.13 and 3.14).

Examine the activities that have been identified on the IFD. Identify any internal pipelines that may be used within activities. Discuss these internal pipelines with those individuals inside the process to decide if the single activities should be replaced with two (or three or four) activities. Mark all pipelines between these activities.

For each pipeline, you should ask the following questions:

- What components (that is, packages from other pipelines) or activities, if any, are needed to transform this item?

- Where do the components come from?

- Are the identified incoming pipelines to be transformed into any of the components or are other incoming pipelines necessary?

- If the identified pipelines are transformed into any of the components, what intervening activities are required to effect the transformation?

Enter files on the IFD to represent any repositories that the user discusses. Identify the contents of each file so as to identify the pipelines in and out of it.

If necessary, modify the boundary of the process being mapped. An unidentified input may be a key component of one of the pipelines. Further, if an incoming

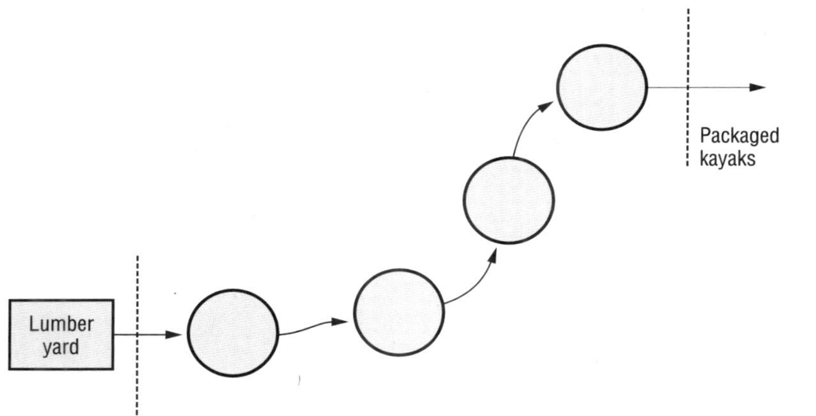

Figure 3.12 Main flow of materials.

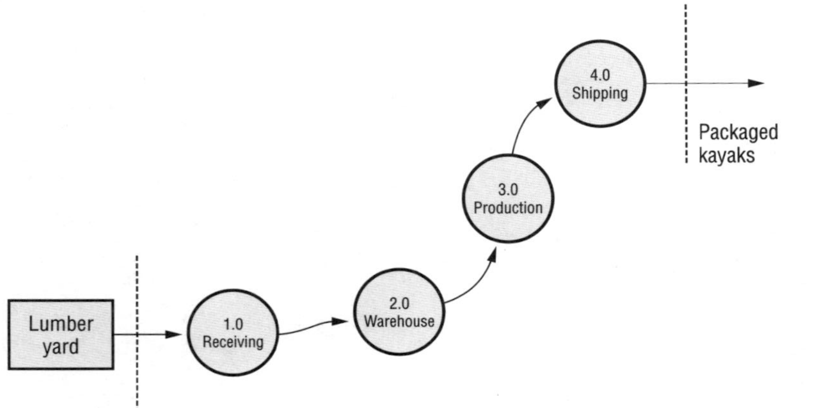

Figure 3.13   Named activities in flow of materials.

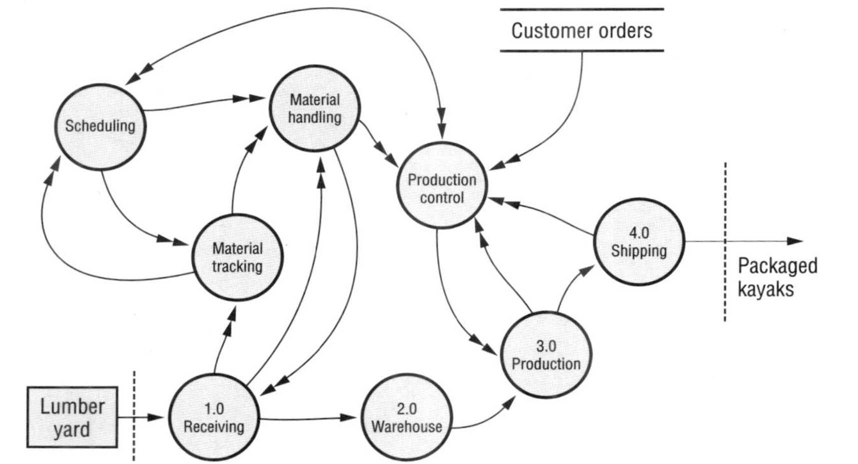

Figure 3.14 Mapped process.

pipeline vanishes and it serves no use inside the context, then remove it. A situation may develop where the scope of the process includes not one but two disconnected networks—one of which is outside the domain of the study. In such a case, eliminate the network that is not being studied from the diagram.

Suggestions for naming activities are:

- Make sure the activity names are honest and that they reflect the activity. For example, an activity name of "painting" is not appropriate if the activity paints a part, drills holes in the part, and inserts a second part into the drilled holes.

- Activity names should consist of a single strong action verb and a singular object. If there are two verbs, further partitioning should be considered.

- Avoid unnameable activities. Divide the process into two or three activities, or group several activities, to identify a process that can be easily named.

3. Label the Pipelines

The names selected for pipelines greatly affect the clarity of the IFD. The following are suggestions for naming pipelines:

- Avoid names such as "data" and "information."

- Be careful not to group disparate items together into one pipeline.

- No two pipelines should have the same name.

- Names should represent not only the package, but also what is known about the package.

Pipelines that move into and out of files do not require names—the file name will suffice to describe the pipeline. All other pipelines must be named.

Case Study

While collecting the flow of information within the plant, Premier Folding Kayaks' project team identified a series of communications that occurred within the production process that did not show up within the business inter-action model. Some of these communications include move location tickets, material ID numbers, allocation requests, and others. These were noted and used to create the IFD in Figure 3.15.

4. Start Over

Mapping a process with an IFD is not easy. False starts and iterations will be needed to appropriately map a process with the IFD approach. You will rarely map the process correctly the first time, until you have experience in developing IFDs. Typically, a team's subsequent IFDs are vastly improved over earlier versions of an IFD for the same process. Earlier versions are usually a learning experience about the process, its pipelines, and its activities. The mechanics of drawing an IFD are trivial. The team should start over any time they feel that significantly better results can be accomplished.

HOW TO DRAW THIS MODEL

The first method to is to use flip-chart paper. Use a pencil. You will be making several changes. The second

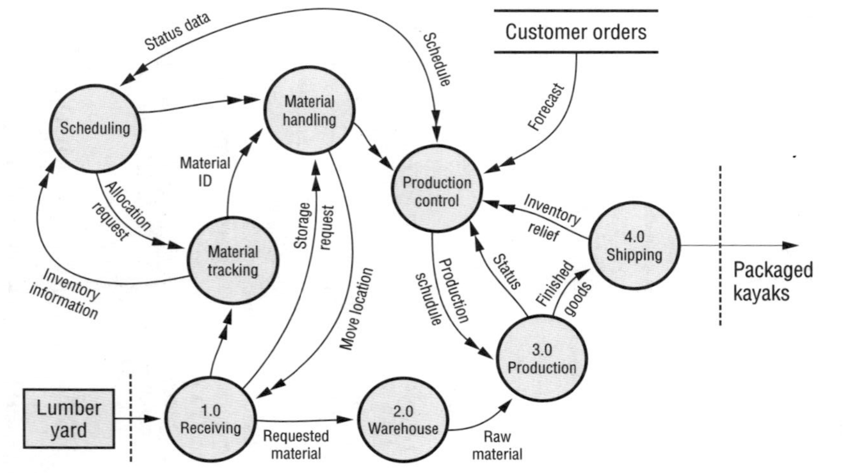

Figure 3.15   Case study diagram.

method is to electronically generate the diagram. There is a variety of software on the market to assist you in drawing an IFD. To gather the data needed you can either use a pad of paper or a PDA.

What to Look for in an IFD

The integrated flow diagram provides you with a different type of picture of the process that you are exploring. The key here is to look for activities that have a large amount of inputs and outputs (that is, five or more). This is usually a sign that there is a great opportunity for defects in this area and should be examined in depth.

SUMMARY

The purpose of this chapter was to outline a second visual tool, the integrated flow diagram. This tool consist of four elements:

1. *Pipelines* show the transfer of verbal or physical items from one spot to another.

2. *Activities* are where work is done.

3. *Files* are temporary holding bins.

4. *External entities* are providers into and out of the diagram but outside the scope of investigation.

Now that you have a map of the process, how do you make changes to it? The next chapter outlines a method to design change.

Chapter 4

Designing Change

In the prior two chapters, you learned about two key visual tools that will help you better understand a process. Now that you have this better understanding, what is next?

This chapter will introduce you to a process for designing change (see Figure 4.1). That is, it will present an approach that you can use to produce new and challenging alternatives that will address the issues/problems/situations identified within your business interaction model and/or integrated flow diagram.

DESIGNING CHANGE OVERVIEW

Step 1: Develop a Transformation Question

After you have reviewed your maps of the process in question, you will typically notice some type of concern—too many inputs or outputs in a specific activity, missing steps, and so on. When there are too many inputs or outputs, the process runs the risk of increased defects. That

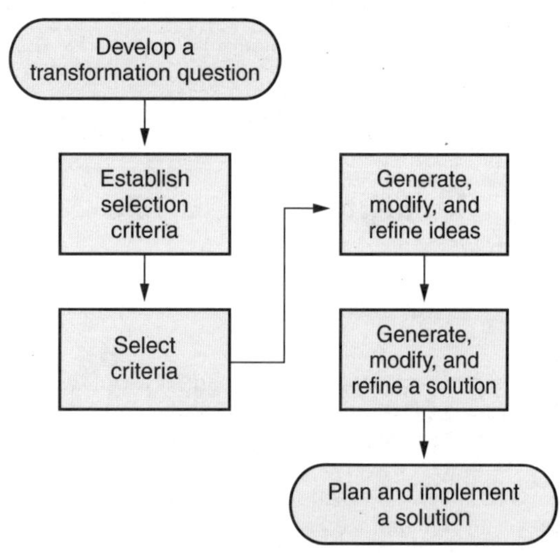

Figure 4.1   Process for designing change.

is, there is a greater opportunity for error. Earlier in the text, it was stated that the goal of a Six Sigma program is to reduce process variation and too many inputs or outputs in a specific activity; missing process steps are typically a sign that a large amount of variation may exist. Focusing on those areas, you may notice a concern or issue that needs to be addressed.

If an issue is identified, the next step is to create a question that will clarify the exact nature and focus of the desired answer you are looking for, for example, to increase the process capability of the sanding area that manufactures the kayak's spars. This transformation question is a description of a concern or problem that requires a solution.

When completed, your transformation question will serve as a guide to developing specific opportunities and solutions, rather than restricting imaginative thoughts due to preconceived notions. Below are a couple of examples of transformation·questions:

- How can we best modify the milling process to increase the process capability to 2.0?

- How can we best increase supplier involvement to reduce incoming inspection?

Step 2: Establish Selection Criteria

Section criteria are used to describe the results expected from a transformation question. It also outlines the constraints that exist in a given situation.

There are two key functions to the step: (1) selection criteria are used to direct and focus idea generation; and (2) selection criteria are used to determine a solution's feasibility and overall performance. To ensure that you are not missing any important selection criteria, it is critical to gain input from all parties affected by the final solution.

Guidelines for Developing Selection Criteria

1. *Selection criteria must be stated in the affirmative.* Since you are trying to explore and develop new solutions, it is important not to use negative terms. For example, the statement "Whatever we do must not interfere with the current production level of our mountain line of kayaks" is negatively

written. A properly written selection criteria would state, "Whatever we do must enhance our production capability of the mountain line of kayaks by a minimum of 50 percent."

2. *Selection criteria should be stated with end results.* The criteria should describe what you want to achieve—not the method for achieving it.

   - End results: Whatever we do must maximize wood usage by leaving no more that 5 percent scrap.

   - Method: Whatever we do must include changes in cutting patterns.

3. *Design criteria should be singular.* Each idea must be independent since it will be expanded in the next few steps. Avoid using the word "and" when writing your criteria.

Step 3: Select Criteria

After you have developed your list of selection criteria, the next step is to review and select those key criteria to expand upon. These are called the "vital few" versus the "trivial many."

Guidelines for Selecting the "Vital Few"

To select the key criteria, review the selection criteria list and choose those which:

- Relate directly to the transformation question

- Can easily generate many ideas

- Do not focus on budget, time, physical, or geographical constraints

It is not unusual to have developed 10 to 20 selection criteria. Upon close review, there may be only one or two criteria that you consider to be the "vital few."

Step 4: Generate, Modify, and Refine Ideas

Generating Ideas

In completing this step, you should use the technique of brainstorming. The four guidelines below will enable you to successfully use this approach.

1. *Defer judgment.* Judgment puts the brakes on thoughts and these brakes need to be removed so that ideas can be generated. During this process, it is best that you write your ideas without concern for their value, feasibility, or significance. Aim for 30 to 50 ideas per person.

2. *Be wild and crazy.* When writing down your ideas, any idea is possible. There is no right answer at this point. Nothing is too outlandish at this phase. Products such as the helicopter, the space shuttle design, and a variety of other commercial and consumer products are a result of "being wild and crazy" when brainstorming. Let your imagination be your guide.

3. *Generate as many ideas as possible.* Research has found that going for quantity is a productive aspect of idea generation. Most people can generate at least five ways of doing any single task. In order to develop 30 to 50 ideas, it is essential that you go beyond the obvious by exploring "the wild and crazy side."

4. *Build and improve ideas.* This guideline encourages you to listen to the ideas generated by the other members of your team. Combine two or more ideas into a single idea, build upon an idea with another idea, and improve upon other ideas. For example, if someone writes "blue" you can say "blue-red" or "blue-X." Have no limits here; go for quantity.

Modify and Refine Ideas

The final solution is usually based upon a combination of several good ideas rather than a single idea. Group your ideas into categories or themes, and then select the best ideas from each category.

1. *Group ideas according to themes.* When reviewing the ideas generated, there are usually at least three or four broad themes. Once you have identified them, list each idea under its respective heading.

2. *Select ideas according to their rated desirability.* Review each idea in each theme and rate each idea as either high, medium, or low according to the selection criteria below. Table 4.1 displays a table that can help you rate and evaluate each idea.

Table 4.1   Rating and evaluating ideas.

Idea	Ease of Use	Innovation	Acceptability	Timing	Rating
Idea #1	Med	High	High	Low	
Idea #2	Low	Low	Med	Med	

- *Ease of use:* If this idea is used, how easy will it be to implement?

- *Innovative:* Does the idea provide a solution which is new or fresh—or is it a similar variation of a current idea presently in use?

- *Acceptability:* If the idea is used, will it be acceptable to those it will affect?

- *Timing:* How long will it take to develop the idea in order to implement it? Do other factors need to be present before the idea can be implemented or can it be implemented now?

Please keep in mind that you are selecting ideas according to how much they appeal to you. This process is not intended to be a rigid set of rules. It is only an aid for final selection.

Guidelines for Creating Proposed Solutions

The "selected ideas" are usually in a rough or outline form. To be useful, each idea should be expanded into a complete statement to show what it would look like in the final solution. Expand each idea by answering the following questions:

1. How would this idea be implemented?

2. How many people, resources, materials, and so on would be involved?

3. What time frame should be included?

4. How do I develop the idea so that it is understood by other people?

After you have expanded each idea, you need to explore if you can combine them into a proposed solution.

Step 5: Generate, Modify, and Refine a Solution

In the prior steps, you were asked to defer judgment and evaluation of the ideas. This was critical for developing the broadest and deepest range of ideas. In this step, you will conduct a critical analysis in order to determine feasibility.

Generate a Solution

1. *Rate each idea.* Using Table 4.1 for each idea, assign a score of 1 to any low rating, a score of 3 to any medium rating, and a score of 9 to any high rating. Sum the total score for each idea and place it in the rating column. See Table 4.2.

2. *Identify critical selection criteria.* Review all criteria identified in step 3. Identify those criteria that you and/or your team deems to be critical. A critical selection criteria is defined as an absolute requirement. These critical criteria are your protection devices, they insure that in the "rush of creativity" you do not violate existing boundaries.

Table 4.2   Scoring the ideas.

Idea	Ease of Use	Innovation	Acceptability	Timing	Rating
Idea #1	Med	High	High	Low	22
Idea #2	Low	Low	Med	Med	6

When reviewing your selection criteria, select those that meet the following guidelines: (1) they must be measurable; and (2) there must be no ambiguity in determining whether an idea does or does not satisfy a critical criteria.

3. *Identify desirable selection criteria.* From the same list of criteria defined in step 3, identify those criteria that you deem to be desirable. Desirable criteria are those criteria that are not designated as limits. They are not used as absolutes to judge ideas, but rather are used to assess relative fit. Take each desirable criteria and assess each idea.

Modify and Refine the Final Solution

This substep takes all the material generated in the prior activity and creates a final solution. That is, it involves a modification or fine-tuning of the proposed solution based on a series of guidelines outlined in the prior sections. These modifications may be necessary in order to create the final solution. Two types of modifications can be made:

- Deleting from the proposed solutions any idea that does not fit a critical selection criteria or poorly fits a desirable criteria.

- Revising an idea so that it meets a critical selection criteria or better fits a desirable criteria.

SUMMARY

At this point of the model, you have completed a cycle through which you have developed and decided on a solution to improve your process. It consists of the seven major steps shown in Figure 4.1, page 54.

The final activity ultimately determines success or failure—that of implementing your solution.

Chapter 5

Closing Comments

One can find many definitions of Six Sigma. The one that I like to use is: "the reduction of process variation by the reduction of defect rate and/or cycle time." Within this definition, it is implied that one's process is defined and can be analyzed.

As stated earlier in this guide, when attempting to define a process, most practitioners use standard flowcharting. Flowcharting is a good tool and is helpful by showing decision points. It does not, however, show departmental interactions or communication patterns. That is where this pocket guide fits in. It provides you with two additional visual tools that give you a big-picture view of your process.

Business interaction diagrams provide you with a picture of interactions that reside in your organization and detailed step-by-step actions (workflow models). The integrated flow diagram provides you with communications flows. Together, these mapping tools provide necessary information to make effective and profitable change.

The tools are not hard to use, but are very powerful. Have fun on your journey in defining and analyzing your processes.

Reference Listing

O ver the past several years, I have completed a list of articles and books that have proven signifi- cantly helpful in the area of process modeling and analysis. Below is this listing. Please note that there are other highly qualified references and support documents that exist on this topic. The list below is the material I commonly use.

REFERENCES

Adair, C. B., and B. A. Murray. *Breakthrough Process Redesign: New Pathways to Customer Value.* New York: AMACOM, 1994.

Allee, V. "Tools for Reengineering." *Executive Excellence* (February 1995): 17–18.

Allender, H. D., "Is Reengineering Compatible with Total Quality Management?" *Industrial Engineering* (September 1994): 41–44.

Andrews, D. C., and S. K. Stalick. *Business Reengineering: The Survival Guide.* Englewood Cliffs, NJ: Yourdon Press/Prentice-Hall, 1994.

Armistead, C. "Principles of Business Process Management." *Managing Service Quality* 6, no. 6 (1996): 48–52.

Asaithambi, D. "Reengineering at Houston Lighting & Power Company." *TMA Journal* (March/April 1995): 12–18.

Atwater, J. B. "Implementing Quality Improvement Programs Using the Focusing Steps of the Theory of Constraints." *International Journal of Technology Management* 16, no. 4/5/6 (1998).

Bennett, J. "Current Issues in Business Process Reengineering." *International Journal of Operations & Production Management* 15, no. 11 (1995): 37–52.

Bennis, W. G., and M. Mische. *The 21st Century Organization: Reinventing through Reengineering.* San Diego, CA: Pfeiffer & Company, 1995.

Brunet, S., and H. Gardin. *Pratiques du Reengineering: Redessine-moi l'entreprise.* Paris: ESF iditeur, 1995.

Burke, G., and J. Peppard. *Examining Business Process Reengineering: Current Perspectives and Research Directions.* London: Kogan Page, 1995.

Carr, D. K., K. J. Hard, and W. J. Trahant. *Managing the Change Process: A Field Book for Change Agents, Consultants, Team Leaders, and Reengineering Managers.* New York: McGraw-Hill, 1996.

Champy, J., and N. Nohria. *Fast Forward: The Best Ideas on Managing Business Change.* Boston: Harvard Business School Press, 1996.

Champy, J. *Reengineering Management: The Mandate for New Leadership.* New York: Harper Business, 1995.

Currid C. et al. *Computing Strategies for Reengineering Your Organization.* Rocklin, CA: Prima Publishing, 1994.

Cleland, D. I. *Strategic Management of Teams.* New York: John Wiley & Sons, 1996

Cole, R. E., "Reengineering the Corporation: A Review Essay." *Quality Management Journal* (July 1994): 77–85.

Cotter, J. J. *The 20% Solution: Using Rapid Redesign to Create Tomorrow's Organizations Today.* New York: John Wiley & Sons, 1995.

Cox, J. F. III. "The Research Agenda for a New Business Philosophy." *APICS-TPA* 7, no.12 (1997): http://lionhrtpub.com/apics/apics-12-97/Cox.html.

Currid, C. et al. *The Reengineering Toolkit: 15 Tools and Technologies for Reengineering Your Organization.* Rocklin, CA: Prima Publishing, 1994.

Davenport, T. H. *Process Innovation: Reengineering Work through Innovation Technology.* Boston: Harvard Business School Press, 1993.

Dekins, E., and H. H. Makgill. "What Killed BPR? Some Evidence from the Literature." *Business Process Management Journal* 3, no. 1 (1995): 81–107.

Dettmer, H. W. "Theory of Constraints: A System-Level Approach to Continuous Improvement." http://www.rogo.com/cac/dettmer1.html (1996): 1–4.

Dixon, J. R., P. Arnold, J. Heineke, J. S. Kim, and P. Mulligan. "Business Process Reengineering: Improving in New Strategic Directions." *California Management Review* (summer 1994): 93–108.

Donovan, J. J. *Business Reengineering with Information Technology: Sustaining Your Business Advantage: An Implementation Guide.* Englewood Cliffs, NJ: PTR Prentice Hall, 1994.

Farrell, J. "A Practical Guide for Implementing Reengineering." *Planning Review* (March/April 1994): 40–45.

Fowler, A. "Operations Management and Systematic Modeling As Frameworks for BPR." *International Journal of Operations & Production Management* 18, no. 9/10 (1998): 1028–56.

Gardiner, S. C. "The Evolution of the Theory of Constraints." *Industrial Management* (May/June 1994): 13–15.

Gattenio, C. A. "Finance Transformation: Making It Work." *Journal of Accountancy* (January 1997): 52,55.

Gewirtz, D. *The Flexible Enterprise: How to Reinvent Your Company, Unlock Your Strengths and Prosper in a Changing World.* New York: John Wiley & Sons, 1996.

Guimaraes, T. "Empirically Assessing the Impact of BPR on Manufacturing Firms," *International Journal of Operations & Production Management* 16, no. 8 (1996): 5–28.

Hammer, M. "Reengineering Work: Don't Automate, Obliterate." *Harvard Business Review* (July/August 1990): 104–12.

Hammer, M., and S. A. Stanton. *The Reengineering Revolution: A Handbook.* New York: Harper Business, 1995.

Hammer, M., and J. Champy. "The Promise of Reengineering." *Fortune* (May 3, 1993): 94–97.

Harbour, J. L. *The Process Reengineering Workbook: Practical Steps to Working Faster and Smarter through Process Improvement.* New York: Quality Resources, 1994.

Harmon, R. L. *Reinventing the Business: Preparing Today's Enterprise for Tomorrow's Technology.* Foreword by L. D. Peterson. New York: Free Press, 1996.

Harrington, H. J. "Continuous versus Breakthrough Improvement: Finding the Right Answer." *Business Process Management Journal* 3, no. 1 (1995): 31–49.

———. "The New Model for Improvement—Total Improvement Management." *Business Process Reengineering & Management Journal* 1, no. 1 (1995): 31–43.

Harrison, A. "Business Process Reengineering: Lessons from Operations Management." *International Journal of Operations & Production Management* 15, no. 12 (1995): 46–58.

Holmstrom, J. "Reengineering in Sales and Distribution— Creating a Flexible and Integrated Operation." *Business Process Reengineering & Management Journal* 2, no. 2 (1996): 23–38.

Homa, P. "Business Process Reengineering." *Business Process Reengineering & Management Journal* 1, no. 3 (1995): 10–30.

Hovav, A. "Adapting Business Process Redesign Concepts to Learning Processes." *Business Process Management Journal* 4, no. 3 (1998): 186–203.

Hunt, D. V. *Reengineering: Leveraging the Power of Integrated Product Development.* Essex Junction, VT: Oliver Wright Publications, 1993.

———. *Process Mapping: How to Reengineer Your Business Processes.* New York: John Wiley & Sons, 1996.

———. *The Survival Factor: An Action Guide to Immproving Your Business Today.* Essex Junction, VT: Omneo/Oliver Wright Publications, 1994.

Institute of Industrial Engineers. *Business Process Reengineering: Current Issues and Applications.* Industrial Engineering and Management Press, Institute of Industrial Engineers, 1993.

Johansson, H. J. *Business Process Reengineering: Breakpoint Strategies for Market.* New York: John Wiley & Sons, 1993.

Juran, J. M. *Managerial Breakthrough: The Classic Book on Improving Management Performance.* New York: McGraw-Hill, 1995.

Karon, P. "Hallmark Welcomes Change in Handling Finances." *Infoworld* (December 19, 1994): 62.

Kelada, J. N. "Is Reengineering Replacing Total Quality?" *Quality Progress* (December 1994): 79–83.

Kinni, T. B. "Get Smart, Reengineer." *Industry Week* (February 20, 1995): 50–52.

Klein, M. M. "10 Principles of Reengineering." *Executive Excellence* (February 1995): 20.

Lau, R. S. M. "Business Reengineering: The Ultimate Productivity Gain." *South Dakota Business Review* (September 1993): 1–5.

Lee, R. G. "Business Process Management: A Review and Evaluation." *Business Process Management Journal* 4, no. 3 (1998): 214–25.

Leth, S. A. "Critical Success Factors for Reengineering Business Processes." *National Productivity Review* (autumn 1994): 557–68.

Love, P. E. D. "From BPR to CPR—Conceptualising Reengineering in Construction." *Business Process Management Journal* 4, no. 4 (1998): 291–305.

Lowenthal, J. N. *Reengineering the Organization: A Step-by-Step Approach to Corporate Revitalization.* Milwaukee: ASQC Quality Press, 1994.

Lucas, H. C. *The T-Form Organization: Using Technology to Design Organizations for the 21st Century.* San Francisco: Jossey-Bass Publishers, 1996.

Machin, S. "Implications of Business Process Management for Operations Management." *International Journal of Operations & Production Management* 17, no. 9 (1998): 886–98.

Mackness, J. R. "Teaching the Meaning of Manufacturing Synchronisation Using Simple Simulation Models." *International Journal of Operations & Production Management* 18, no. 3 (1998): 246–59.

Manganelli, R. L., and M. M. Klein. *The Reengineering Handbook: A Step-by-Step Guide to Business Transformation.* New York: AMACOM, 1994.

Mariotti, J. "The Antidote for Reengineering." *Industry Week* (April 15, 1996): 20.

McAdam, R. "An Integrated Business Improvement Methodology to Refocus Business Improvement Efforts." *Business Process Reengineering & Management Journal* 2, no. 1 (1996): 63–71.

McKenzie, J. *Paradox—The Next Strategic Dimension: Using Conflict to Reenergize Your Business.* New York: McGraw-Hill, 1996.

McQueen, R. J. "Product Flow, Breadth, and Complexity of Business Processes." *Business Process Reengineering & Management Journal* 2, no. 2 (1996): 8–22.

Moskal, B. S. "Reengineering without Downsizing." *Industry Week* (February 19, 1996): 23–28.

Motwani, J. "Business Process Reengineering." *International Journal of Operations & Production Management* 18, no. 9/10 (1998): 964–77.

Murphy, D. "How One Healthcare Provider Is Meeting the Challenges of the '90s through Reengineering." *TMA Journal* (March/April 1995): 19–22.

National Institute of Standards and Technology (USA). United States Department of Commerce. National Technical Information Service. *Integration Definition for Function Modeling (IDEFO).* Gaithersburg, MD: National Institute of Standards and Technology, 1993.

Ould, M. A. *Business Processes: Modelling and Analysis for Reengineering and Improvement.* New York: John Wiley & Sons, 1995.

Peel, D. "Causes and Impact of Reengineering." *Business Process Management Journal* 4, no. 1 (1998): 44–55.

Peppard, J., and P. Rowland. *The Essence of Business Process Reengineering.* Englewood Cliffs, NJ: Prentice Hall, 1995.

Perez, J. L. "TOC for World Class Global Supply Chain Management." *Computers & Industrial Engineering* 33, no. 1–2 (1997): 289–93.

Petrozzo, D. P., and J. C. Stepper. *Successful Reengineering.* New York: Van Nostrand Reinhold, 1994.

Radnor, Z. "A Characterization of a Business Process." *International Journal of Operations & Production Management* 18, no. 9/10 (1998): 924–36.

Recardo, R. J., and D. J. Jones. "A Report Card on Reengineering." *Production and Inventory Management Journal* (Third Quarter 1997).

Rizzo, T. "Corporate Rightsizing and the Theory of Constraints."
    http://www.rogo.com/cac/rizzo1.html (1994): 1–4.
———. "The Corporate Team and the Theory of Constraints."
    http://www.rogo.com/cac/rizzo2.html (1994): 1–2.
———. "The Theory of Constraints."
    http://www.rogo.com/cac/rizzo11.html (1994): 1–5.
Roberts, L. *Process Reengineering: The Key to Achieving
    Breakthrough Success.* Milwaukee: ASQC Quality
    Press, 1994.
Sandoval, V. *Les Techniques du Reengineering.* Paris:
    Hermhs, 1994.
Smith, J. J. "Theory of Constraints and MRP II: From Theory
    to Results." http://www.rogo.com/cac/JJSmith.html (1996):
    1–16.
Soliman, F. "Optimum Level of Process Mapping and Least
    Cost Business Process Reengineering." *International
    Journal of Operations & Production Management* 18,
    no. 9/10 (1998): 810–16.
Strub, M. Z. "Quality at Warp Speed: Reengineering at AT&T."
    Bulletin of the American Society for Information Science
    (April/May 1994): 17–19.
Swartz, J. B. *The Hunters and the Hunted: A Nonlinear
    Solution for Reengineering the Workplace.* Portland,
    OR: Productivity Press, 1994.
Taylor, B. *Successful Change Strategies: Chief Executives in
    Action.* Hemel Hempstead: Director Books, 1994.
The Economist Intelligence Unit, Andersen Consulting.
    *Business Reengineering in Asia: Transforming
    Organisations to Meet the Global Challenge.* London:
    The Economist Intelligence Unit, 1995.
Tinnila, M. "Strategic Perspective to Business Process
    Redesign." *Business Process Reengineering &
    Management Journal* 1, no. 1 (1995): 44–59.
Tobin, D. R. *Transformational Learning: Renewing Your
    Company through Knowledge and Skills.* New York:
    John Wiley & Sons, 1996.

Tomasko, R. M. *Go for Growth!: Five Paths to Profit and Success—Choose the Right One for You and Your Company.* New York: John Wiley & Sons, 1996.

Towers, S. *Business Process Reengineering: A Practical Handbook for Executives.* Cheltenham, UK: Stanley Thornes, 1994.

Van Meel, J. W., and H. G. Sol. "Business Engineering: Dynamic Modeling Instruments for a Dynamic World." *Simulation & Gaming* (December 1996): 440–61.

Wellins, R. S., W. C. Byham, and G. R. Dixon. *Inside Teams: How 20 World-Class Organizations Are Winning through Teamwork.* San Francisco: Jossey-Bass, 1994.

Wilkinson, R. "Reengineering: Industrial Engineering in Action." *Industrial Engineering* (August 1991): 47–49.

Wright, D. T. "Software Tools Supporting Business Process Analysis and Modeling." *Business Process Management Journal* 3, no. 2 (1997): 133–50.

Zairi, M. "Business Process Reengineering and Management." *Business Process Reengineering & Management Journal* 1, no. 1 (1995): 8–30.

———. "Effective Process Management through Performance Measurement." *Business Process Reengineering & Management Journal* 1, no. 1 (1995): 75–88.

———. "Best Practice in the Car After-Sales Service: An Empirical Study of Ford, Toyota, Nissan, and Fiat in Germany—Part 1." *Business Process Reengineering & Management Journal* 2, no. 2 (1996): 39–56.

Index

Note: Italicized page numbers indicate illustrations.

A

B

D

designing change
    establish selection criteria, 55–56
    ideas, 57–60
    select criteria, 56–57
    solutions, 60–61
    transformation question, 53–55

E

events, 8–9
external entities, 39–40

F

file, 38–39
final solution, modifying/refining, 61
flowcharting, ix , 40, *41,* 63
functions, 3, 20, 22

G

generating ideas, 57–58
generating solutions, 60–61
goal model, 13–14

H

hierarchy models, 11–16

I

ideas, 57–59
information events, 8
input/output pipelines, 42, *43*
integrated flow diagram (IFD)
    drawing guidelines, 40, 42–49
    drawing the model, 49–51
    elements of, 33–40
    leveled, 40
    structured analysis, 31–32
interrelated models, 11, *12*

J

L

M

O

P